*"THE AUTOMOBILE AND
 ITS GOOD
  AIR CONDITIONING"*

-Israel Mustelier-

LICENSES.

AC Type: Universal.

AC Type: Automotive.

ACKNOWLEDGMENTS.

Certificate of Completion

Pinellas Technical Education Centers

This certifies that

Israel Mustelier

has completed 300 hours and Occupational Completion Point A in the

Electricity Technology

under the standards approved by the Florida Department of Education Division of Workforce Development
Given this month of September 2006

Site Administrator
Clearwater Campus

Diploma for completing the electricity course.

Letter for completing the course for automotive mechanic.

Stratford Career Institute
12 Champlain Commons, PO Box 1560
St. Albans, VT 05478-5560
Tel.: 1-800-435-5338

Office At:
8675 Darnley, Mount Royal, QC H4T 1X2

******************AUTO**MIXED AADC 054

Israel Mustelier 9538 T7 1872
4021 12th Ave N
Saint Petersburg FL 33713-5911

Student No.: F312244
Assignment No.: AEC5-D
Date Graded: October 27, 2009
Reference No.: PAL200910276

Overall Grade: 90%

Dear Israel Mustelier,

I have reviewed your work for: **Auto Mechanics**

It is with great pleasure that I take this opportunity to congratulate you on the successful completion of your program of studies. You have worked hard, showing the determination to take on every challenge, and the perseverance to keep at it until your goal has been met.

I have always known that successful home study students are a special breed of individual. Most work on their own, with no need for someone standing over their shoulder telling them what to do. Some come home from a long day of work, only to spend another couple of hours studying before turning in for the night. Others have faced unemployment and uncertainty, determined to come out on top at the end of a long, hard struggle with suitable employment in a career of their own choosing. Whatever your own situation has been, you are among an elite group of dedicated individuals.

As at any other school, there are some who enroll in our courses who fall by the wayside before graduating. You have had what it takes to finish what you have started, and I congratulate you. I know that your present or future employer will appreciate what your diploma says about you, not only in terms of the knowledge and skills you have gained, but the personal qualities you have demonstrated in getting there.

Your diploma will be sent out to you separately, in special packaging, once you have completed your tuition payments. It is school policy that the tuition must be settled before we can issue the diploma. The enclosed statement indicates the number of monthly payments still to be made, and also the lump sum amount, should you wish to cover the remaining balance in one payment. Doing so will allow you to receive your diploma shortly thereafter.

Once again, please accept my heart-felt congratulations and best wishes to you for continued success.

Sincerely,

Claude Major, PhD
Vice-President and
Director of Education

To all decided:

The air conditioning system in cars, like any other system, play a big role. It is true that some vehicles could pass through the broken air, but it would be annoying and unpleasant, especially if there is heat. For some people living in countries developed, a good car with a non-functioning AC system is compared to an individual with an excellent suit, but with shoes in poor condition.

We will see here how the AC system works and some of the breaks that are more common and how to repair them. Not all vehicles will carry out the repair the same; as the techniques may differ, it is not my intention to fix a rule nor do I intend to be accused of proving erroneous data. Rather, this is only a representation of how to do the work by our own means, thus saving our money given to a well-intentioned person who would at the moment be offered to do the work.

Index

I- THE AUTOMOBILE AND ITS FUEL   -   9

II- THE AIR CONDITIONING SYSTEM   -   20

III- MAIN TOOLS   -   32

IV- REFRIGERATING   -   48

V- THE ELECTRICAL SYSTEM   -   53

VI- REPAIRS IN THE AC SYSTEM   -   68

VII- HEATING AND COOLING SYSTEM   -   76

I

THE AUTOMOBILE AND ITS FUEL

Although the purpose is to focus on the air conditioning system and then on the heating, any individual who is skilled in repairing automotive AC systems, he should know some basic aspects of how the vehicles move. For example: although a cardiologist is an expert in his branch, must study a generality of the human body. From before entering into the matter, we will stop at the fuel system.

THE FUEL SYSTEM consists of the following elements:

1- The fuel tank.

2- The fuel pump.

3- The filter.

4- Fuel lines.

5- Fuel pressure regulator.

6- Fuel rail assembly.

7- The injectors.

BRIEF DESCRIPTION OF THE SYSTEM

Let's start with "the gas pump". This component is a small machine that, through a pipe, sends gas from the tank to the engine of the car. Measuring the gas pressure, we could determine how the pump is working. This is usually be located inside the tank. But in some cars, it may be out of the tank.

Using as an example the current Chrysler and Dodge: its gas tank is one 16-gallon capacity and is located at the rear of the vehicle; below it. In then, as in many other brands, the gas pump is located inside the tank. By its part, the gasoline filter does exactly what its name indicates: filters impurities separating them from gasoline. Because of high fuel prices, many people lately they never fill the tank. If you often drive with the tank almost empty, the moisture in the air inside the tank can cause an effect that contaminates the gasoline. As result: the filter may end up cramping, and possibly even the injectors are affected. The fuel filter (in the line that takes the fuel to the engine) can be seen in a Plymouth Voyager by the bottom of one of the doors; in other car brands, the location differs.

There is another very important component: the fuel pressure regulator. This can be found in the motor zone; where the fuel rail assembly is located.

The fuel rail assembly or injector rail, apart from picking up the injectors, can carry the fuel pressure regulator. This small element maintains an appropriate gas pressure and puts fuel back to the tank through another line known as "return". While brands like Toyota and Nissan, have the pressure regulator on this site: in the nozzle rail; on the contrary, the Dodge models have them inside the tank of gasoline, mounted on the fuel pump. But there is a fact: although the location of the fuel pressure regulator is not the same in all cars, the function is the same.

The injectors, one for each cylinder, are attached at the beginning to the fuel rail assembly and in its end look at an intake hole in the car's engine.

(Fig. 1) -Fuel pump.

(Fig. 2) -Gasoline filter.

(Fig, 3) -Fuel pressure regulator of a Toyota Corolla

(Fig. 4) -The fuel pressure regulator attached to the end of the fuel rail assembly. This species of the tube has the injectors that, like syringes, introduce gasoline to the engine.

FREQUENTLY ASKED QUESTIONS

I- The engine of the car tries to work, but does not start.

a- There is no fuel pressure. Maybe the car ran out of gas.

b- The fuel line is blocked by someplace, perhaps disconnected.

c- The gas pump has failed.

d- The petrol pump relay is faulty.

II- There is a long attempt to get the engine to boot. Finally, start.

a- There may be very low fuel pressure. Check the fuel pressure regulator.

b- The nozzles may leak.

c- The ventilation of the gas tank may be blocked. Remove tank cap and try to boot like this.

d- The problem may be at the gas pump.

III- Once the engine is started and the car starts to move, there is poor performance. Sometimes the engine is turned off.

a- Low fuel pressure. Maybe the pump is failing.

b- The fuel filter is dense. Needs replacement.

c- Dense injectors. Add to the fuel tank "injectors cleaner" solution.

CHECKING THE FUEL PUMP AND PRESSURE.

If everything indicates that the fuel pressure is low, first inspect the lines; so you will see that the issue is not just a gasoline leak. Then remove the tank cap and tell an assistant to move the starter key to the on position. For about seconds you should hear the sound of the pump. The car starts and the sound must be continuous, although difficult to perceive. If there is no sound, the pump is not responding.

Check the relay fuel pump. Can be under the instrument panel, near the column of address. Maybe it's under the hood, in the relay box. Or could even be close to the gas pump, next to the tank itself. Check that the relay is reaching 12 volts. Also, check the fuses in the passenger area. Check the electrical connector at the petrol pump.

1- Remove the Schrader valve cap. This valve is in the fuel rail assembly (in some 4 cylinder engines).

2- Attach the pressure gauge to the Schrader valve.

3- In 6-cylinder engines there is probably no Schrader valve. In all cases where this service port does not exist, a special adaptation (T-shape) can be installed between the line supplies gasoline and fuel rail assembly. For that:

A-disconnect the supply line from the fuel rail assembly.

B-install the T between the supply line and the fuel rail assembly.

C-install the pressure gauge to the T.

4- Start the engine and observe the reading given by the meter. Maybe between 35 and 45 psi is OK.

5- Wait fifteen minutes. If the pressure is kept low, perhaps at 0 psi, check for some restriction on the gasoline line. If the line is OK, the filter may be dense or the pump failing.

(Fig. 5) -Pressure Meter.

(Fig. 6) -Special fitting that can be placed between the gas supply line and the fuel rail assembly.

(Fig. 7) -The connection point between the supply line and the fuel rail assembly. Note the ring of closing.

(Fig. 8) -The fuel rail assembly. It has on the left edge the fuel pressure regulator. For its right edge comes the gas supply line; in this connection the T.

REPLACING THE FUEL PUMP

1- Disconnect the negative battery cable.

2- Empty the fuel from the tank. You can remove the plug that is in the background and if not, you can extract the fuel by means of a siphon pump.

3- Raise the car with some safe jacks.

4- Remove the gas tank from the vehicle.

5- Find the gas pump and remove it from its place. You must also detach the connector electric.

Note: If you put a new pump, the best option is to install a new gasoline filter.

THE INJECTORS.

CHECKING IT:

The following method for checking the injectors is done with the auto-off engine.

a- Disconnect the electrical connector of the first injector.

b- With the electrical connector disconnected, use a jumper wire to connect an injector terminal to the positive terminal of the battery.

c- Attach another wire jumper to the other end of the injector. Quickly connects and disconnects to the negative terminal of the battery. The injector should make a slight sound. If it does not make sound some, the injector is not good. Should be replaced.

d- Continue to examine the other injectors in this way.

The next method to check the injectors is the engine on, without being accelerated:

a- Disconnect the electrical connector of the first injector.

In doing so, it should be appreciated that momentarily a decrease of the engine rpm. Then increase the rpm again.

b- If this does not happen and the motor continues at a constant rhythm of rpm, it is shown that the injector is not working as it should.

c- Do the same with the following injectors.

REPLACING THEM.

1- Disconnect the negative battery cable.

2- If something gets in the way, remove it to gain access to the fuel rail assembly.

3- Separate the gas supply line from the fuel rail assembly.

4- Disconnect the electrical connectors from the injectors.

5- Remove the fuel rail assembly from the vehicle. To do this, remove the bolts. Then withdraw the fuel rail assembly with the injectors attached.

6- Using a screwdriver, remove the clip securing the injector to the fuel rail assembly. Releases the fuel rail assembly.

(Fig. 9) -Injectors.

II

THE AIR CONDITIONING SYSTEM

THE AC SYSTEM consists of the following elements:

1. The compressor.

2- The condenser.

3- The Accumulator. Note: If the system does not have an accumulator, it may there is a receiver-dryer.

4- The expansion valve. Note: If the system does not have an expansion valve, an orifice tube may be there in its place.

5- The evaporator.

These five elements communicate with one another by means of hoses or pipes. The coolant circulates and there is the whole system AC of the cars. How simple!

But to better understand this extraordinary system, take a look at the following figures. Then you will not find it difficult to identify each item in your car.

(Fig.10) -Compressor.

It is next to the engine. Carries a pulley. Some cars have them visible at the top of the engine. While others from the bottom; being thus more difficult to access the time to arrive in order to replace him.

(Fig. 11) -Condenser.

It is in front of the car. It resembles a radiator and is located in front of it.

(Fig. 12) -Accumulator.

While this is almost always located at the exit line of the evaporator; that is, on the Low side of the system. On the contrary, the receiver-dryer is usually located on the High side of the system. Or what is the same: in the line of entry to the evaporator. Remember that generally, the vehicle carries only one of them; receiver-dryer or accumulator.

Fig. 13) -Receiver-dryer.

Unlike the accumulator, the receiver dryer has on the top a small peephole showing the refrigerant circulating in the system.

(Fig. 14) -Orifice tube.

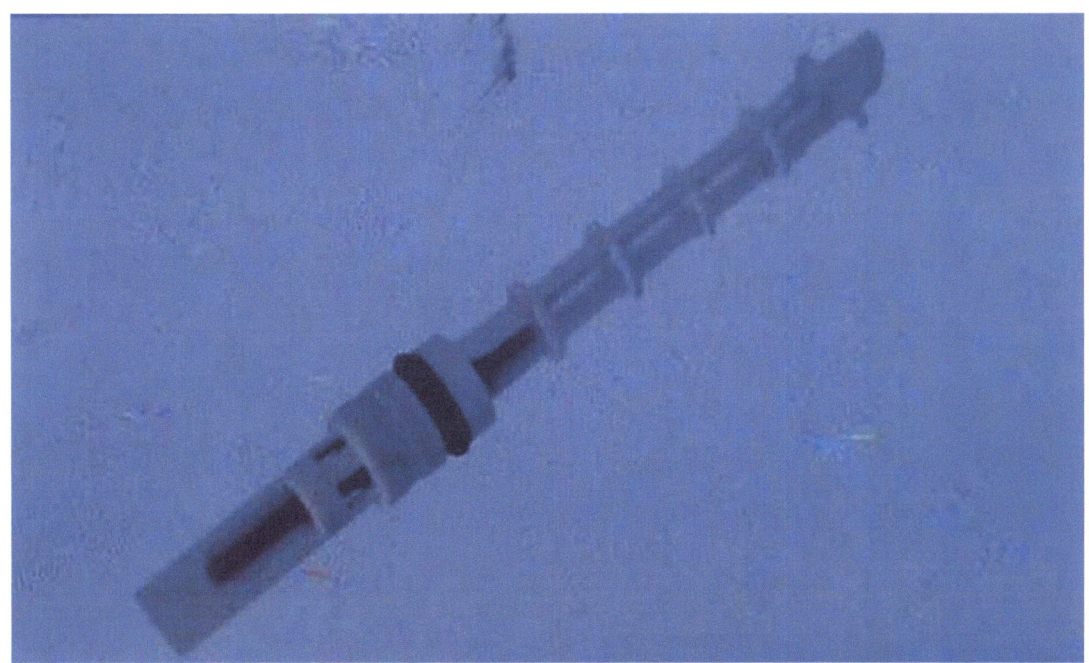

(Fig. 15) -Expansion valve.

The expansion valve is well attached to the evaporator. Also, the orifice tube is located close to the evaporator, in the entry line to the same or high side of the system. Remember that usually, the vehicle carries only one of them: orifice tube or expansion valve.

(Fig. 16) -Evaporator.

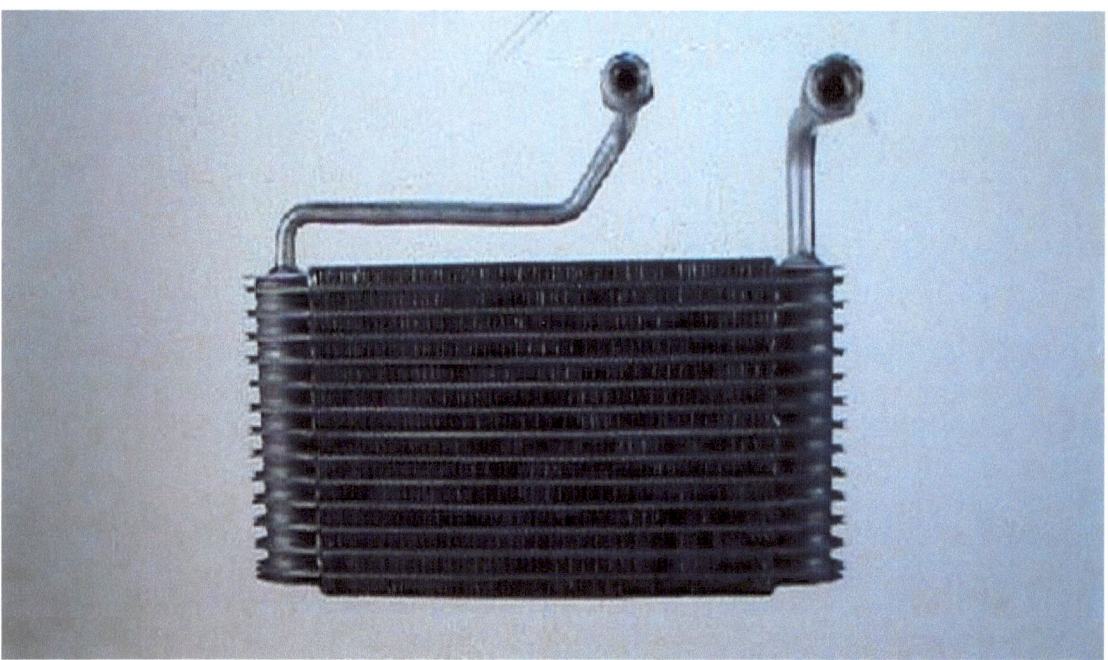

Almost always under the instrument panel (in the passenger area) some cars have them. The evaporator is next to another component called heater core. And both were they will be tucked inside a housing or wrapping. Also, there will be in the same housing a blower.

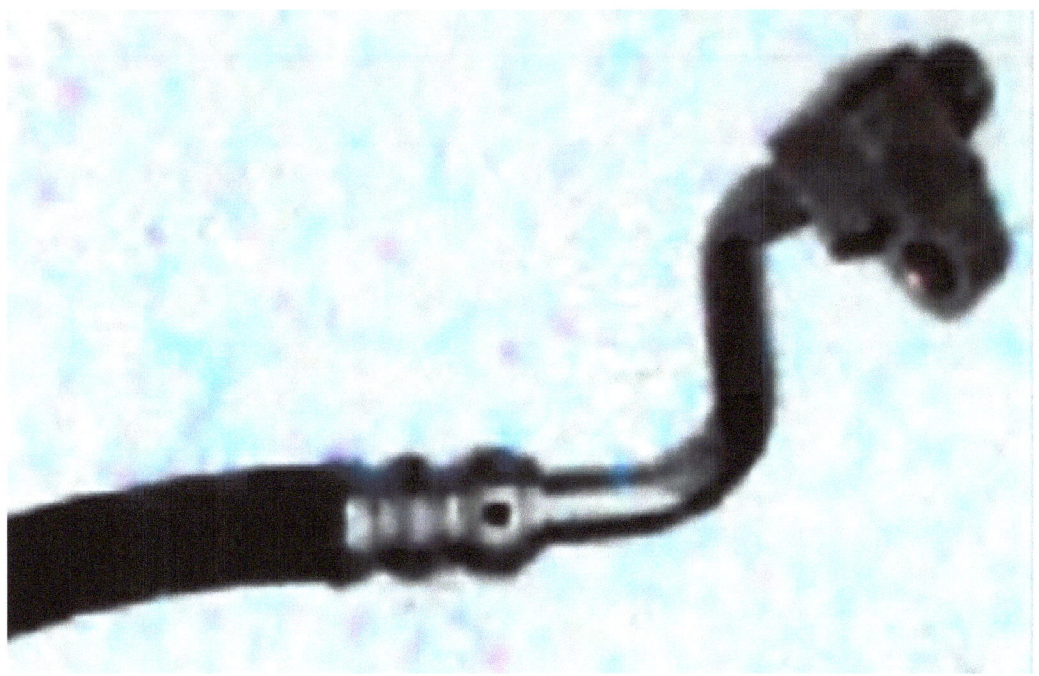

(Fig. 17) -Hose.

Through hoses like this, the components that make up the AC system are joined. They are the conduits to circulate the gas.

THE CYCLE

In this book, we will briefly discuss various automotive systems with their respective cycles. And although it does not have to do mechanics with anatomy, I would like to mention the circulatory system of man. The reason is that the AC system keeps a slight similarity to the human circulatory system. The latter is composed of a heart that through rhythmic contractions, circulates the blood through a pipe that does not end: the arteries and veins. While the arteries go to the organs of the body, the veins come thereof. Thus each organ will receive the blood, oxygen from the lungs. This precious liquid is purified by the kidneys. After the organs assimilate the oxygen, the blood will remove the waste and lead to the lungs again, to be released abroad. The cycle is always the same.

Something similar happens in the AC system.

1- We have the compressor first. This, like the heart, is the bomb that makes circulate the gas through the uninterrupted piping of the system. For the time in which it is written this book, cars use a refrigerant called R134a. The compressor adsorbs by a side a low pressure of this gas, with low temperature. This side we will call it "side of suction". The gas is pumped and exits on the other side of the compressor with a high temperature and pressure. To this other side, we will identify him like "side of discharge".

2-The condenser. Take the hot gas comes as steam from the compressor. This vapor refrigerant enters the condenser cap and flows through the winding. As flows, going down, the temperature of the gas goes down. It is good to note that the condenser receives fresh air from a fan located before it. Finally, the gas that has entered as vapor, it will become liquid. And so, as a liquid, the gas will leave the condenser through the down.

3- The dryer filter: it serves as a container or deposit for the liquid gas that has left the condenser. The dryer filter will knowingly give the evaporator what it needs, since the evaporator demands a varied sum of gas, according to the conditions of use.

Protects the entire AC system, how? Containing inside a drying agent that absorbs the moisture of the gas.

Note: The function of the accumulator corresponds to that of the receiver-dryer. That is, it stores gas and removes moisture. Moisture is an impurity that would bring the entire AC system to an inefficient or non-functioning state. So that the accumulator, or its equivalent, has been the kidney of this system. Makes the refrigerant pure.

If any liquid gas leaks out from the evaporator, it will be stored in the accumulator. The liquid gas would damage the compressor if it were not for the intervention of the small important tank we are talking about.

4- The expansion valve: In order for the cooling in the passenger area to be the correct, the amount of gas towards the evaporator must be controlled. This will be achieved through the expansion valve. A bore that opens and closes the valve, changes the coolant arrival pressure from high to low pressure. So that the evaporator will receive the appropriate amount of gas, at the discretion of the person operating the AC.

5- The evaporator: When we turn on the air conditioning in our vehicle, an air hot from the passenger area will hit the evaporator. This component receives likewise the refrigerant supplied by the expansion valve. This gas comes cold and liquid. And while the gas passes through the winding of the evaporator, the warm air moves above said winding. As the liquid gas receives heat from this air, a change of state takes place (from a low liquid pressure to a low vapor pressure). The temperature of the gas in the form of the outlet of the evaporator must be higher than the temperature of the gas in liquid form at the inlet of the evaporator.

Warm air blown on the evaporator contains some moisture. The humidity will be condensed in the evaporator and drained out of the water. A drain pipe in the evaporator bottom will drive this water away from the car.

An important component of the evaporator to cool down is the blower. This fan is as already mentioned in the evaporator area. It is said fan who attracts to the hot air from the passenger compartment, on the evaporator. Also, blows the cooled air passing through the evaporator, outwardly. That is, towards the passenger area.

Then I will present the refrigerant cycle in a rustic way so that it has a vision of how the gas flows in these AC systems. This cycle, as in the circulatory system of living beings, is invariably repeated. But before, it is good to keep in mind that in all air conditioning systems, there are two sides: one high side and one side of low (low side).

The high side is the part that transports gas high pressure and high temperature. This identified with the service port whose cover is usually red. It starts at the exit of the compressor (discharge side) and continues through the condenser. Ends at expansion valve (or the orifice tube).

The low side is the other part of the system. The gas has low pressure and its temperature is low. Start at the exit of the expansion valve and continue through the evaporator. Ends at the inlet of the compressor. It is identified with the service port whose cover is usually blue; sometimes it is black.

In a particular part of the hoses or AC pipes that are in your car, you can locate the two service ports. The pipe on the low side is thicker than the on the high side.

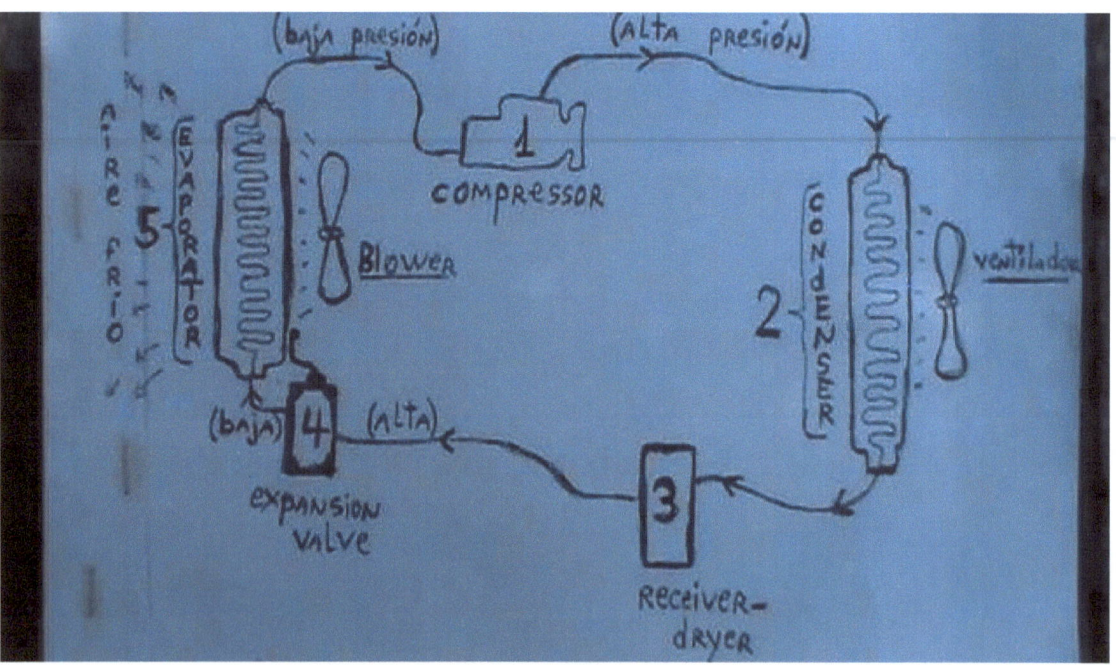

(Fig. 18) -Schema of an automotive AC system.

Course taught by Israel Mustelier

(Fig. 19) -The two service ports of the AC system in a Toyota.

III

MAIN TOOLS

An important factor to perform a good repair in the AC system is the use of the appropriate tools. The manifold (or set of meters) is indispensable. As well the use of the vacuum pump and a recovery machine plays a significant role. Too much say, you will need screwdrivers, tweezers, and wrenches. An air compressor will be of great utility, like good jacks and ramps to raise the vehicle. With a thermometer, you can measure the temperature in the grate coming from the evaporator. No AC technician should lack a multimeter because sometimes the problem is electric. For example, the compressor is not reaching 12 volts when it performs its function. Or maybe the fan condenser is not powered. There is also a tool that will not always be used, but at a time may be useful. If it is necessary to replace the orifice tube, the following key can help:

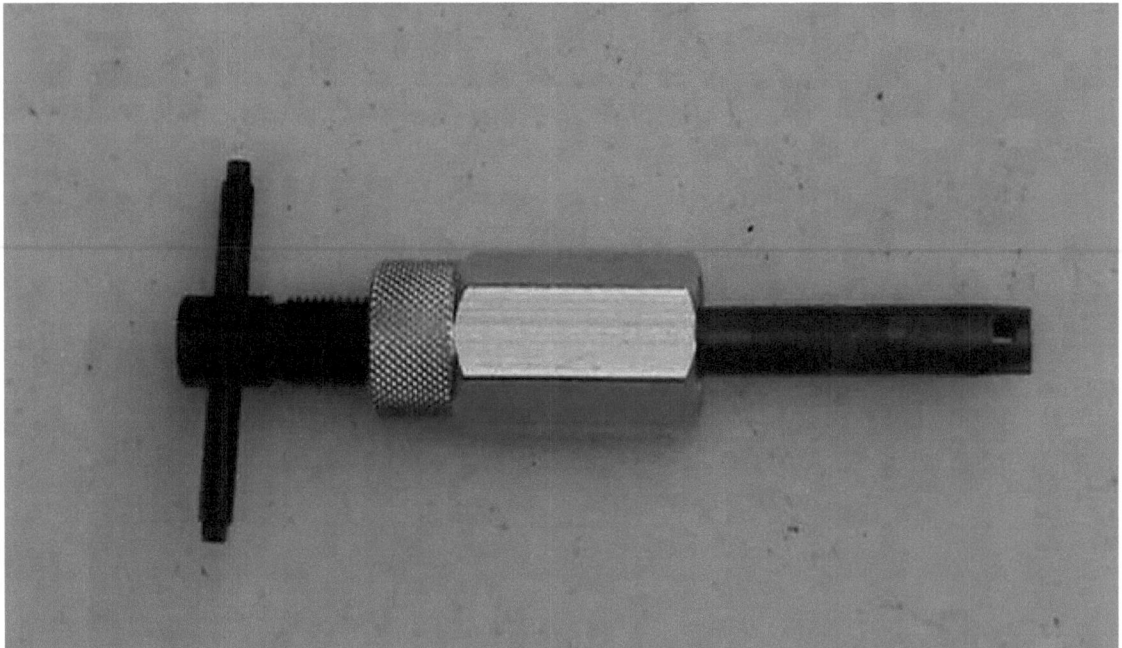

(Fig. 20) -Orifice tube tool.

(Fig. 21) -Multimeter.

(Fig. 22) -Ramps.

In this office, it is wonderful to have an electronic leak detector, as in addition some ultraviolet glasses and a UV flashlight. And something important: always uses the means of security necessary to avoid accidents. Wear glasses to protect your eyes, perhaps gloves or aprons. Avoid working with garments: they may become entangled and cause damage not only to the vehicle, also to you. If your hair is long, pick it up before you begin.

(Fig. 23) -Manifold gauge set.

I- The <u>manifold</u> or set of meters: Without doubts is the most important tool for an AC technician. The sphere with the blue color is used to measure the refrigerant pressure on the side of low. Its scale reaches up to 150 psi (pounds per square inch). The sphere with red color is used to measure the high side. Its scale reaches 500 psi.

When measuring with the set of meters, the two valves must be closed. The clockwise, with the valves closed, will mark the pressure in the refrigerant reigns in the system. Something to keep in mind: never open the high side valve while the AC system is running. This could break the valve and even cause damage to your person. In addition, the set of meters you will use for an automotive system differs which is used for a commercial or residential system. So keep in mind that you should use R134a meters for the vehicles.

The meters have three hoses: one blue, one yellow and the third is red. The blue hose is screwed under the blue watch of the game and tied in the low service port that it can find in the car. The red hose is screwed under the game's red clock and attached to the service port on the high side of the system. The yellow hose is screwed into the middle of the meter and serves as the refrigerant to the system by means of a full bottle that has been purchased before start work.

Next, we will stay in the readings obtained with the game of meters. But before, there are certain factors that must be known.

psi is the "atmospheric pressure". This is 14.7

The pressure reading on the clocks varies according to the height above sea level. Is of about a half pound less per thousand feet above sea level. For example: to zero feet above sea level, the atmospheric pressure will be 14.7 psi. But 1000 feet above the sea level, the atmospheric pressure will be about 14.2 psi.

Also, the pressure reading on the manifold clocks varies by temperature environmental. That is, the higher the temperature in the environment, the greater the reading on the clocks.

The automobile and its good air conditioning

To start using the manifold, do the following:

1- Install the manifold or set of meters to the high and low service ports of your car. The couplings are made to hold firmly in ports avoiding gas leakage.

2- Start the engine of the vehicle.

3- Turn on the air conditioner and set it to the maximum.

4- Observe the readings given by the manifold for the low-side and the high-side.

5- Compare these readings with the existing "Table for Refrigerant R134a".

I- Reading: Low side -normal. High side -normal. (The system is working fine).

II- Reading: Low side -low. High side -low. (The system has little refrigerant).

III- Reading: Low side -low. High side -high. (There is a gas lock, perhaps in the expansion valve).

**

IV- Reading: Low side -high. High side -low. (Compressor is failing).

V - Reading: Low side -high. High side -high. (There is an overload of refrigerant in the system).

TABLE FOR REFRIGERANT R134^a.

Room temperature In degrees Fahrenheit (F°)	Low side (PSI)	High side (PSI)
65	25 - 35	135 - 155
70	35 - 40	145 - 160
75	35 - 45	150 - 170
80	40 - 50	175 - 210
85	45 - 55	225 - 250
90	45 - 55	250 - 270
95	50 - 55	275 - 300
100	50 - 55	315 - 325
105	50 - 55	330 - 335

If the AC system has little refrigerant, add up to a required amount. If the expansion valve locks the system, replace it. If the compressor is not working, remove and install a new one. And if you see a refrigerant overload in the system, eliminate it a little until you reach the limit suitable. Important: If you detect that the system does not have any refrigerant, the most logical is conclude that there is a leak. Locate, repair and refill.

(Fig. 24) -Coupling set for low and high service ports. The hoses in the manifold are curled up to them.

(Fig. 25) -Thermometer.

II- The thermometer: Used to measure the temperature. To put it to operate: press the trigger button. Press the small back button to activate the beam projected at the point you want to measure. The thermometer display will tell the temperature.

(Fig. 26) -Vacuum pump.

III- The <u>vacuum pump:</u> If an AC system has suffered a significant loss of refrigerant (Perhaps empty) and one or more of the elements the system it has stopped working properly, the system must be "evacuated" before adding new refrigerant. The moisture and air must be extracted from the entire system by the vacuum pump. This step of extracting is an "evacuation". This operation is important because humidity and air are contaminants that are not compatible with the refrigerant. Obstruct their movement through the system and create corrosive acids. The refrigerant that has been contaminated must be extracted urgent.

The vacuum pump scale ranges from 0 to 30 in-Hg (inches of mercury). The reading to indicate a total vacuum varies by height above sea level. The standard reading of zero feet above sea level is 29.92 in-Hg.

HOW TO USE THE VACUUM PUMP

1- Connect the blue manifold hose to the low service port of the AC system.

2- Connect the red manifold hose to the discharge service port.

3- Connect the yellow hose from the center of the manifold to the vacuum pump socket. Open the discharge valve of the vacuum pump or remove the discharge cap, in case of taking it.

4- Open the two manifold valves: red and blue. Start the vacuum pump.

5- When you spend some time, with the red clock set to 0, blue will read 28 or 29 in-Hg. A nap cannot be reached, turn off the vacuum pump and look for a possible drain in the hoses or in the unions. Then turn the pump on again and keep it to function with the aforesaid number for half an hour.

6- After this time, turn off the vacuum pump and close the red manifold valve. The pressure may rise slightly, it is normal.

7- Look at the blue manifold clock. Make sure that the vacuum level stays the same for about five minutes. If it stays the same, it shows that there is no leak. But if by otherwise, vacuum loss is indicated, there is leakage in the system. This problem needs to be fixed; if not, when the refrigerant is added, it will be lost little by little ends to zero again.

8- Close the blue valve.

9- No moisture in the system and no exhaust samples, everything will be ready to fill with the new refrigerant.

10- Disconnect the yellow hose from the vacuum pump and then connect it to the bottle of the refrigerant to be placed in the AC system.

11- Turn on the vehicle and also your air conditioning. Opens the blue valve on the manifold.

Keep the red closed.

12- Fill the system as compared to the table to R134a. You must know what temperature environment exists. To do this use the thermometer. When the manifold reading is appropriate, stop the recharge.

13- Arrive in the cabin of the passengers so that you notice the good freshness that reigns. Make sure to hear the sound of the evaporator fan. It must be working. Using the thermometer measures the temperature in one of the windows leading to the evaporator. Between 40 and 43°F comes from wonders.

(Fig. 27) -Electronic leak detector.

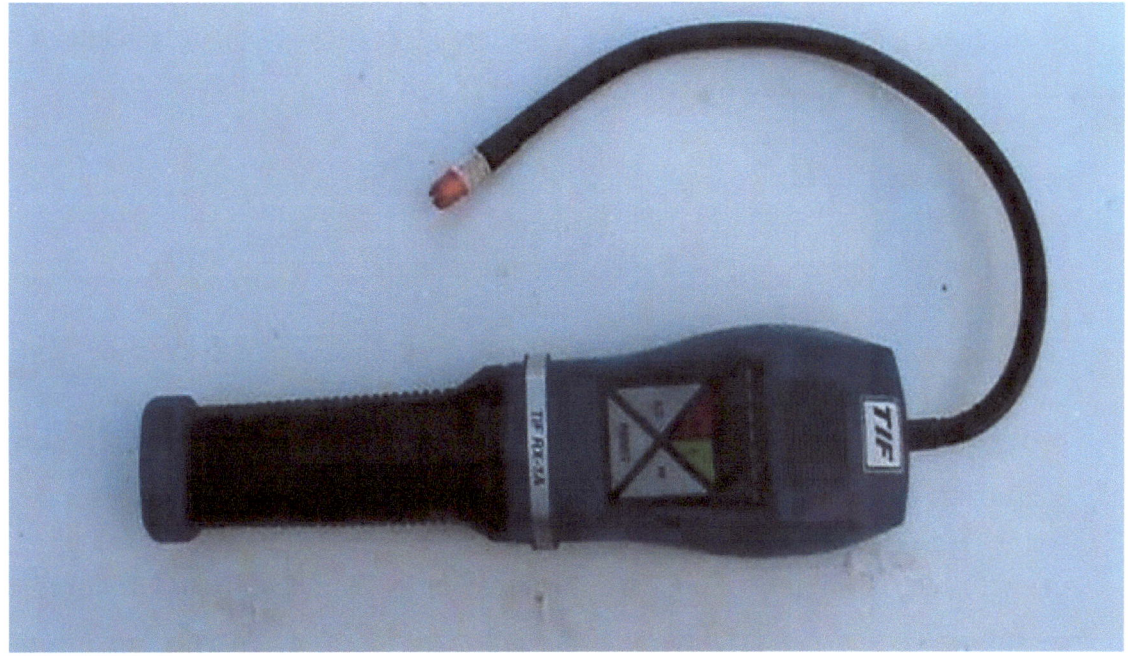

IV- <u>Leak detector</u>: When there is a low level of refrigerant in the system, it must be assumed that there is a leak. It should detect where it is in order to repair it. The leaks can occur due to loose connections, damaged hoses or lines damaged metal. Many leaks result from normal vibration, which loosens thread connections or joins metal fatigue, causing breakage. To repair these leak needs to be repaired (or perhaps replace) the affected area.

FINDING LEAKS

There are several options:

- "Option soapy water": Apply soapy water on the alleged escape area. With a toothbrush, applies to all pipe joints and connections. Start the engine the car and turn on the AC. The escaping refrigerant will cause the soapy water solution bubbles occur. Once detected the leak, tighten with two wrenches loose union or repairs leaking components, maybe the compressor or perhaps the condenser. Or maybe escape it is in the worm of one of the valves in the system. Squeezes the worm with a little key of worms. If evil continues, replace it.

-- "Option ultraviolet (UV)": a UV glasses are needed, a UV flashlight, a bottle of UV dye to detect leaks and a bottle of refrigerant R134a.

1- Start the engine and air conditioning.

2- Use the UV glasses. Add the dye UV low valve.

3- Then add R134a refrigerant, also low valve. This is done with the purpose of distributing the dye UV throughout the system.

4- Turn off the AC and turn off the engine.

5- With the flashlight looking carefully leaks by unions, by worms of valves, for the compressor's body, by the accumulator, through the condenser. To check the evaporator, water drained see for symptoms of contamination.

6- Having found the leak, fix.

---"Option electronic leak detector":

Start the engine and AC system. Turn on the leak detector by pressing the power button. The lights will illuminate during startup. With the high light on and a repeated sound: pi, pi, pi, pi;

the detector is ready to find leaks. Softly, walking with the tool, placing its sensor near all system components and joins. The sensor hose is handy and can bend or stay right at your whim. Once the instrument detected the leak, it will emit a different tone and the light signal will increase.

(Fig. 28) -Compressor.

V- The <u>air compressor:</u> The refrigerant which has been contaminated within a system of AC it must be removed. It should never be recharged with new refrigerant putting with old refrigerant contaminated. The new one also is unclean. So that the AC system must be downloaded, completely emptied the old refrigerant. It should evacuate or remove air and moisture from the vacuum pump. And then, in the game will enter a new process. Although it is not always necessary: the system should "rinse". This is where comes the use of the air compressor. But... what is rinse? It is to remove dirt internal system: foreign matter like the air and the moisture impair the proper functioning of the mechanism. Rinsing is essential whenever there is a breakage compressor replacement. If not rinsed, certain metal particles that came from the old compressor could circulate again and return to the new compressor to cause it a serious injury or even leave it running. Thus, the new compressor will last a short time.

FLUSHING A SYSTEM.

1- (Depending on the model gun rinse) Connect a terminal hose that is used to the gun rinse. The other terminal of the hose to the air compressor. The cylinder or container from the gun must contain liquid "clean and flush" to be used.

2- It is good to note that it can rinse any of the system components: a hose, the condenser, the evaporator. The latter two do not have to be removed from the vehicle to be rinsed. And it can also clean the entire system. But it should not rinse the compressor or the accumulator alone. Doing so would remove the oil and could damage the internal parts of either.

To rinse part of the system, it can say a line or hose, put it into position vertical. Fixed a terminal rinse hose on the input of the line. The output of the line inside a container placed underneath. Insert the muzzle of the gun to the other hose terminal. Press the trigger control. Start the air compressor. The liquid is pressurized by the air compressor, then clean the line and also go down to the container below.

To flush the entire system, first, remove the expansion valve or if be bears, orifice tube. Unscrew the cylinder cap rinse and put enough liquid "clean and flush" inside. Screw the cylinder cap. With the air compressor walk, pressurizes cylinder between 90 and 125 psi. Because this cleaning removes oil, it is necessary to replace it. For the condenser: about 2 ounces. For the evaporator: about 2 ounces. For the receiver-dryer: 1 ounce. For the accumulator: 4 ounces. For the compressor: 4 ounces. Note that the oil in these cases, it can be PAG refrigerant 46, 100 or 150. Check with a person trained or use the manufacturer's manual to find out what kind of oil takes your car.

(Fig. 29) -Flusher gun.

(Fig. 30) -Recovery machine.

VI- The recovery machine: If you need to remove or download refrigerant from the system, it can be stored using a recovery machine. This refrigerant is not lost. Rather, it could be used again in the future. Usually, the recovery possesses a high side where the red hose is connected and a low side where it will be connected the blue hose. Open valves from recovery machine it will allow refrigerant coming out and finally enter inside one recovery tank by using the machine.

If it were to use the stored refrigerant, make sure it is the type you need. Never mix different refrigerants in a system. This could cause damage. Neither the refrigerant R22 or its successor R410a are used for automobiles.

R134a it is only used on automotive systems. Or perhaps, a rare case, the refrigerant R12.

Note: Since 1992, the R12 was behind because of the damage to the ozone layer.

(Fig. 31) -Liquid "clean and flush". Excellent for rinsing AC systems.

(Fig. 32) -12 oz can of refrigerant.

IV

REFRIGERATING

An old saying goes: -the ignorant are like people that do not see-. There is nothing so true! Sometimes something comes easy in the AC system and we left the success thinking that is not the opportune time to make expenditures, or perhaps quickly we go to someone to solve the problem. We are willing to pay what he asks of us. Surely this is an expense unnecessary because we can either fix it ourselves.

For example, the vehicle cooled slightly. As an immediate solution is added refrigerant. You do not have to call anyone unless that person wants to do him a favor. Neither you have to wait long. Simply do the following:

- Acquires the refrigerant R134a to need. You can choose the can with 12 ounces, but the bottle has better conditions. The next illustration shows an ideal container for a person who has never done this operation, but this person can make it effective. Note that the bottle came with its hose and with its coupling. Also, bring a small clock whose needle to indicate when the system is full.

The automobile and its good air conditioning

(Fig. 33) -Bottle of R134^a and can of 12oz.

1- Find in your car the low service port. Remember that almost always has a blue cap and is on the thickest refrigerant line. Remove the cap.

2- Connect the coupling of the bottle to the service port low. The coupling comes like a ring to finger.

3- Start the engine. Turn on the air conditioning and set it at the maximum.

4- During recharging, hold the bottle vertically with the top down. Every two or three seconds, rotate it between 12:00 and 3:00 hours of a clock. Shake constantly from one side to another.

5- Follow the process, checking the pressure indicated on the clock of the bottle.

6- When you done, remove the docking port. Place the blue cap where before and... enjoy the air!

It is also easy to fill the system with the can of 12 oz. You can do so:

1- Take in your hands a valve refrigerant can open. Turn its handle from right to left, to the maximum.

2- Screw this valve on the can of 12 oz.

3- Take in your hands the manifold and make sure that the two valves are closed.

4- Screw the yellow hose of the manifold on the valve refrigerant can open.

5- Focus your attention now on the low service port of the vehicle. Remove the cap protective. Connect the blue hose coupling of the manifold over it.

6- Open the blue valve manifold and opens the valve refrigerant can open. First turn its handle from left to right to pierce the seal of the can. Then you give laps in the opposite direction to open and give out the refrigerant. Start the engine of the car. Turn on the AC and set it at the maximum.

7- As same is done with the bottle of R134a, place the can of 12 oz vertically with the top downwards and raised to the height of your face. Shake it while charging the system.

8- Compare the blue clock reading manifold with the table for refrigerant R134a. Fill the system to the range consistent with the table.

9- When you finish, close the blue valve manifold. Close the valve refrigerant can open. Withdraws the service port manifold of the car and replace the protective cap. Make free the yellow hose from the can.

(Fig. 34) -Valve refrigerant can open.

(Fig. 35 y 36)

-Manifold hoses.

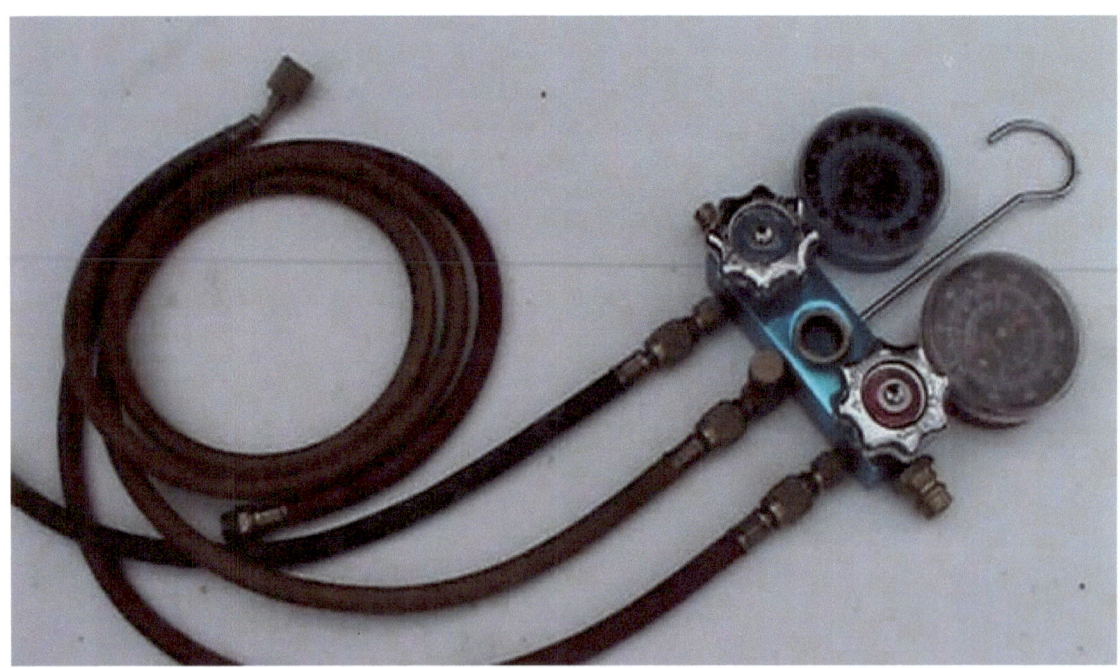

V

THE ELECTRICAL SYSTEM

Since air conditioning depends on electricity to function, we would do well to know something about the electrical system of the cars.

THE ELECTRICAL SYSTEM consists of the following components:

1- Battery.

2- Alternator.

3- Ignition switch.

4- Starter.

5- Ignition coil.

6- Distributor.

7- The spark plugs.

8- The spark plugs wires.

BATTERY: This is the point where the electricity part. Sends 12 volts to various components. The good start of the vehicle depends on a good battery and good electrical cables with good connections. A good starter is indispensable. The major cause of failure of the starter motor is a low voltage battery.

The battery cables are usually low to high caliber, as follow:

4 gauge, 2 gauge, 1 gauge, 2/0 ought, 3/0 ought.

(Fig. 37) -Battery.

PROBLEM: The automobile will not start because of the battery. This one is discharged, or not it makes the starter do its function.

If that happens, you have three options:

1- Check the alternator.

2- Check the cable from the alternator to the battery. It can be loose.

3- Examines the connections (positive and negative) of the battery: the connectors can be loose. Sometimes bring fill good effect to iron out the connectors on its inner circumference. And also, to iron out the battery terminals because eventually could generate a film of dirt. Maybe for you can be incredible, that is real, battery terminals past a time could create an insulator.

THE ALTERNATOR: Its main function is to provide charge to the battery. The alternator's rotor is driven to rotate by the engine pulley vehicle. This important turn creates a field which it produces magnetic alternating current.

CHECKING THE ALTERNATOR:

1- Take the multimeter. First, examines the battery voltage with the engine off. It should be about 12.5 volts.

2- Start the engine and examines the battery again. If the multimeter shows more voltage (13 to 15 volts) the alternator is working.

(Fig. 38) -The alternator.

(Fig. 39) -The pulley rotates the rotor of the alternator and it produces alternating current.

If you need to replace the alternator, do the following:

1- Disconnect the negative battery cable.

2- Remove the alternator pulley by loosening the tension. To do so might loosen required one or two of the alternator bolts and a movement thereof, exposing the pulley (pulley tensioner may be in the pulley of the alternator or may come from another pulley that applies tension to the alternator pulley. Moving said pulley with a wrench adjustable you can remove the tension pulley).

3- Disconnect the electrical cables of the alternator.

4- Remove the bolts securing the alternator to the engine.

5- Remove the alternator. Sometimes you have to force hard to leave.

THE IGNITION SWITCH: The steering column it is the tower which loads the rudder of the automobile. The ignition switch is located on that column and is activated by rotating the key cylinder, that is, the key that stars the vehicle. A symptom of an ignition switch without working is when the engine is stopped completely, the key has no effect; there is not even sound under the hood. You suspect that finally, this would happen because on several occasions gave work starting the engine to turn the cylinder, giving the impression of the existence of a false contact. In this case, you have to replace the ignition switch.

For this you can follow these steps:

1- Disconnect the negative battery cable.

2- If the vehicle it has adjustable steering rudder, you must place it in the lowest position.

3- The tilt steering lever must be removed. Unscrew.

4- Disassemble the top cover and the bottom of the steering column. For that, remove the two Torx screws. Both are on the underside of the cover.

5- Remove the key cylinder. To do this, place the key in the position of roll the engine. Use a screwdriver to disengage the retainer. Pushing out the cylinder.

6- Remove the screws that secure the ignition switch to the steering column. Torx screws should be, perhaps T20.

7- Disconnect the luminous aura and the buzzer on the side of the switch.

8- Disconnect the electrical connections of the switch.

9- Remove the switch from the column.

THE STARTER: For many engines start, they must be rotated. This is the purpose of the small starter: rotate the huge engine. This little component has at its front a grate rotor with teeth which makes an engagement with the teeth engine car. By such gear, to start operating the starter, the vehicle engine starts rotating too.

A defective operating small rotor with teeth can cause the starter is working without achieving the goal: not to rotate the big engine of the automobile. Thus, the starter will make a plaintive sound indicating that the component is operating, that's fine, but there is something: its teeth rotor is not rotating the big engine. The car can not start. The best solution is to replace the entire starter.

Note: If you need to replace the starter, do not hesitate to do so because it is not difficult. First, locate it. It should be at the bottom of the car engine. Use ramps to raise the vehicle and perhaps jacks and towers required. Disconnect the negative cable from the battery. And once the starter is before your face, you have to loosen and remove the bolts that fix it. Remove it from its site when disconnecting the electrical connectors it has.

(Fig. 40) -Starter motor.

THE DISTRIBUTOR: When talking about the distributor, we can say that we have reached the danger zone because we are no longer working with only 12 volts, but with a very high voltage; widely able to kill anyone. Under the hood, there is a transformer known as "ignition coil" which raises the electricity it receives from the battery (12 volts) to a huge sum of voltage (12 000 to 45 000 volts, depending on the type of car). With a cable, the distributor receives from the ignition coil this very high voltage to handle distribution to each spark plug. In many vehicles, a game of cables charging the high voltage. Also know that, according to the number of cylinders in the car, will be the number of spark plugs. For example, a four-cylinder engine has four spark plugs.

(Fig. 41) -Distributor (Note the spark plug wires).

(Fig. 42) -Distributor and spark plug wires in a four-cylinder engine.

Because the high voltage is produced, extreme caution should always be taken that an operation is carried in this area. These components include not only the distributor, the ignition coil, and the spark plug wires; also include items connected as testers of the system. To see how the distributor is running you can do following:

- Have with you a calibrated ignition tester. (See Fig. 43).

- Disconnect the spark plug wire first and install the tester. You must disconnect said a wire on the side of the spark plug, leaving it connected on the side of the distributor.

- The tester clip should be attached to any metal parts in the engine.

- Then, ask an aide for starting the engine without making it fully while you observe the tester. It glows blue, it shows that there is a discharge electric plentiful. To this spark plug, it is coming sufficient current from distributor. If testing the engine stars, the test should not stay more than a minute.

- Continue the test with the second spark plug cable. If the results are as above, there are signs of a good distributor. But one good distributor does not mean that the spark plugs are at its best shinning. They may need replacement to be damaged or failing.

Note: If you decide to change the spark plugs, the best option is also changing wires of these plugs.

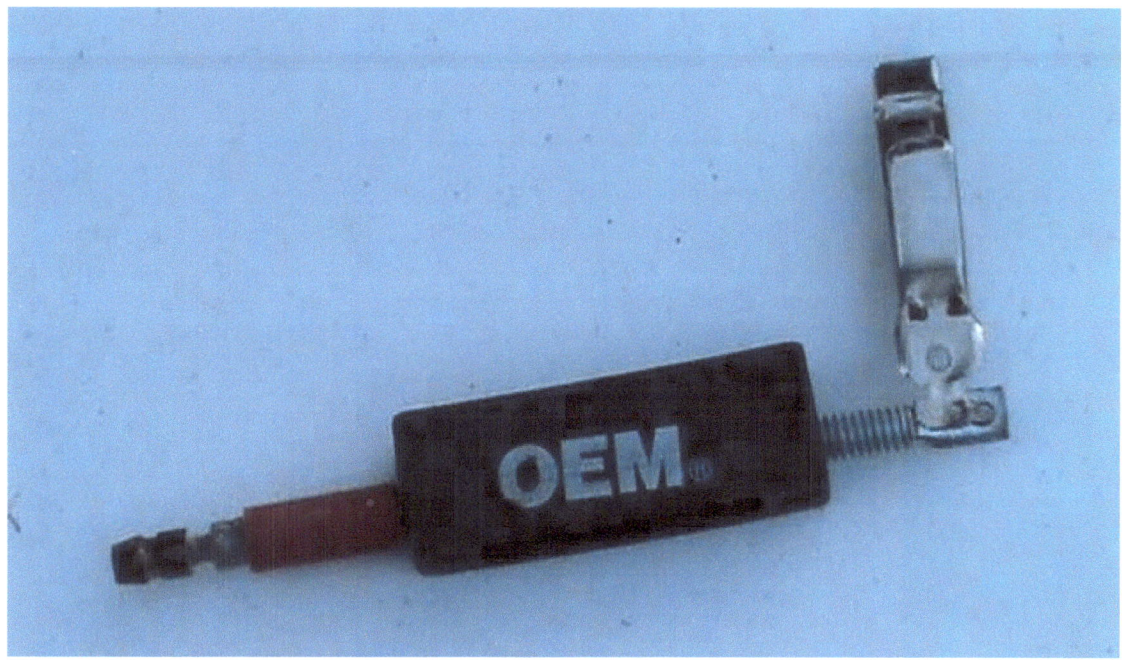

(Fig. 43) -Calibrated ignition tester.

If you need to replace the distributor, do the following:

1- Disconnect the battery cables.

2- If there is anything that hinders to do the job, remove it.

3- Disconnect the electrical connectors of the distributor.

4- Identify each spark plug wire and remove it from the distributor cap.

5- Remove the bracket distributor.

6- Loosen the screws and remove the distributor cap.

7- Remove the nuts securing the distributor and remove it from its site.

(Fig. 44) -Spark plugs.

The company Delco, of "General Motor", it was the maker of the first distributor of power. But as times progress, it has ceased to be used and the ignition with a distributor is replaced by the method of starting the vehicle without a distributor. This system is identified by the initials DIS (Direct Ignition System).

SOME ELECTRICAL PROBLEMS IN THE AC SYSTEM.

*- The condenser fan is good, but it does not receive power.

SOLUTION: Check the fan electrical connector. Or check the fuses in the panel, almost always located on the instrument panel and the driver's side.

*- The compressor is good, but it is not getting its 12 volts to start operating.

SOLUTION: - Check that the power cable compressor has continuity.

- Check the connector that feeds the compressor.

- Check the compressor relay. It is located in the "Power Distribution Center", which is under the hood: in the engine area. Remove the AC Compressor Control Relay from the box. With the engine running and the air conditioning put in on checks voltage. If it is arriving voltage, then is clear that the problem is the relay. Replace it.

(Fig. 45) -Relay compressor.

Note: It is good to clarify that the absence of rotation of the clutch of the compressor can be due to the controls that operate in the system, which light up the compressor and off, due -between other things- to the refrigerant pressure.

The compressor controls:

These controls, as has been said, can turn on and turn off the compressor. The reasons due to several factors: it may be due to a high or low-pressure refrigerant. Another reason is to protect the vehicle from over-cooling. And another reason: to protect the self-compressor. When it is stopped, it is resting overheating that could be lead to burn its winding.

Control: "Low-pressure cut out switch": It is located in the expansion valve or maybe in receiver-dryer. Stops the compressor when the pressure drops due to leakage, blockage or low temperature. In systems using orifice tube, this control is located on the accumulator.

Control: "High -pressure cut out switch": It is located on the refrigerant line. Stops compressor when the system pressure rises due to a blockage or overheating condenser. This switch could also be found in the compressor.

Control: "Super-heat switch": It is located in the rear of the compressor and is exposed to the refrigerant flow. The switch contacts are normally open. But if a predetermined temperature is reached (caused by a decrease of the refrigerant flow) the contacts close.

(Fig. 46) -Relay boxes in the engine area.

VI

REPAIRS IN THE AC SYSTEM

REPLACING THE COMPRESSOR:

If the clutch compressor is not rotating, the compressor itself has not started perform its function. The front cylinder is rotating may under the influence of the pulley, but the clutch is stopped. The clutch compressor is in front of the front cylinder. Reading from the manifold gauge set gives to us right now: low side -high reading; high side -low reading. This is logical because the refrigerant has accumulated on the inlet side of the compressor. But not being it working is not fired by the discharge side.

If the compressor is located on the bottom, almost certainly need to lift the vehicle to do the job. A pair of ramps will be very useful for those moments.

Note: Whenever you replace the compressor, should also replace the accumulator.

1- Remove the negative battery cable.

2- Download the refrigerant from the system.

3- a) Loosen the bolt on the tensioner.

b) Remove screw and tensioner pulley.

c) Remove the compressor pulley.

4-a) Unleash connector that allows it to reach 12 volts to the compressor.

   b) Unleash every existing switch from the compressor.

5- Loosen the bolt that attaches the multiple (hoses) with the compressor. Remove the bolt. Separate multiple from the compressor.

6- Clean with PAG oil the multiple openings. Then cover. You can use paper and tape.

7- Loosen and remove the four bolts that secure the compressor.

8- Remove the compressor instead.

9- Remove the accumulator from its place.

Before installing a new compressor keep in mind the following:

- Add PAG oil required. It can be PAG 46, 100, 150 (about four ounces). Always use the new oil-free moisture. Eliminates the old oil.

- It's good to rotate about ten times the compressor rotor.

- Putting on new o-rings, lubricated with PAG oil.

- Check that the 12 volts will arrive well at the new compressor.

REPLACING THE ACCUMULATOR:

a- The negative battery cable must be disconnected.

b- The refrigerant must be discharged from the system.

c- The compressor must have been removed from its site.

1- Disconnect the input and output lines of the accumulator and cover all openings to avoid contamination and moisture in the system.

2- If the accumulator has pressure sensing switch, remove it. The purpose will be put it on the new accumulator ready to be installed.

3- Remove the bracket from the accumulator. Remove this component from its place.

Before installing the new accumulator, remember the following:

-Put about four ounces of fresh PAG oil inside the new accumulator. (It maybe take less oil, according to the manufacturer's model)

Note: If you were TO REPLACE AC HOSES, note the following:

- When you disconnect each hose, clean both sides of the connection using PAG oil.

- Always use two wrenches when you loosen or tighten hoses.

- Cover the end of the hoses immediately after disconnecting.

REPLACING EXPANSION VALVE

When the expansion valve has stopped its function, it blocks the refrigerant that circulates through the system. The manifold gives the following reading: low reading for low side, high reading for high side. It is logical because the bulk refrigerant is accumulated at the entrance of the broken component; which it is located on the high side.

Some vehicles have the expansion valve next to the firewall, at the rear of the engine area. While others models have it under the instrument panel in the passenger area. Now well, in all cases, the expansion valve is located very closed to the evaporator.

1- Empty the system from the refrigerant.

2- Unleash the line of the expansion valve. This line ends at the evaporator inlet.

3- Remove the expansion valve from its area close to the evaporator. If fitted, remove the connection low- pressure switch.

4- Remove the other line from expansion valve. It is coming from receiver- dryer.

5- Cover the ends of the lines immediately after you have disconnected.

Note: If the expansion valve is under the instrument panel, will require a laborious disassembled of it to reach it.

REPLACING THE ORIFICE TUBE

Never clean and re-install a used orifice tube. Remember that this component is located inside the line that ends in the evaporator inlet.

1- Empty the system from the refrigerant.

2- Disconnect the high side-line at the junction of entering to the evaporator.

3- Put a small amount of PAG oil between the entrance line to orifice tube. The purpose will lubricate the hole.

4- Insert into the line the orifice tube tool. Moves gently the tool to avoid during removal tears of the tube.

5- Lubricates the o-ring of the new orifice tube using clean PAG oil.

6- Insert the new orifice tube in its place between the inlet line to the evaporator. Insert this component with its short end facing the evaporator. Squeeze the orifice tube into the hole.

7- Reconnect the high side line.

Note: **REPLACE THE CONDENSER** is not difficult. As it is located in the front area of the vehicle, front of the radiator; this is a site that is not reach out complicated. Keep in mind when performing the work that you will have the presence of one condenser fan. In necessary case, it will have to remove first. Then, disconnect the refrigerant lines entering and leaving the condenser. Cover the opening soon as possible. Remove the screws holding the condenser and finally remove this component from its site.

REPLACING THE EVAPORATOR

- A leaking gas evaporator must be replaced. If leakage occurs in the joints of refrigerant lines to the same, may be sufficient to tighten the nuts. To examine the leaks in the evaporator, you can do the following. Star the air conditioning with its blower on high speed for about thirty seconds. Then, turn off the air conditioning and fan. Wait for the refrigerant to build up. Insert the sensor of electronic leak detector between hollow evaporator drain, so it does not count water is present. If the detector emits an alarm, the evaporator or line connections almost sure are leaking.

1- Disconnect the negative battery cable.

2- Empty the system from the refrigerant.

3- It is necessary to drain the cooling system of the automobile. In other words: leave the engine dry, without antifreeze coolant. Drain this system by the radiator.

4- Disassemble the instrument panel. This is a big job that requires dedication and a lot of patience.

5- Disconnect the heater core hoses.

6- Disconnect and cover the lines coming to the evaporator.

7- Remove from its site this significant housing containing the evaporator, heater core, and blower.

8- Disconnect the electrical wires coming to said housing.

9- Open the housing and remove the evaporator.

ISSUES TO CONSIDER:

I- The refrigerant pressure is too low on the low side.

Results:

... first) Icing on the evaporator.

... second) Less air circulation through the evaporator.

... third) Warm conditions in the passenger area.

... fourth) Pumping oil. This could damage the compressor valves and if continues the problem, it can burn the compressor.

II- Oil immeasurably in the system.

a) Too much oil will cause oil pumping.

b) Too little oil will cause a rapid deterioration in the compressor bearings, in the pistons, rings, and valves.

(In both cases, the compressor will deteriorate faster)

III- The air conditioning does not cool much.

a) Inspect the condenser. It must be dirty. Clean it up.

b) Check that the compressor turns cloche.

c) Inspect the refrigerant level. If low, it adds.

d) Check the evaporator fan. Check the condenser fan.

e) Inspect the evaporator. If this component is dirty, clean it. And if it clogged, make a flush rinse using a flusher gun and an air compressor.

IV- There is noise in the AC system.

a) Perhaps the compressor pulley is loose.

b) Bolt's compressor needs to be tighter.

c) Compressor bearings are in need of replacement.

d) Oil level in the compressor is low.

e) A damage fan.

V- Water not coming from the system.

a) The drainpipe is clogged.

b) The evaporator fan is not working.

VII

HEATING AND COOLING SYSTEM

The automotive heating and the cooling system touch each other. Or said in metaphor: both shake hands. To better glimpse into the heating world, we would well know what do in the engine the cooling system.

COOLING SYSTEM: consists of the following elements:

1- Water pump.

2- Thermostat.

3- The two radiator hoses (Low hose and high hose).

4- Radiator.

5- Cooling fan.

6- Recovery tank.

This system is filled with "antifreeze", usually a half and a half of water and coolant. Or, anyway, if you're filling an empty cooling system, mix equal parts of coolant and water into a clean container. Some antifreeze comes premixed. Read the label, must say if this the case: 50/50.

The automobile and its good air conditioning

CYCLE

The hot coolant leaves the engine and passes through the thermostat to catch the high hose. Then, enters the radiator found in the front of the vehicle. A fan cools the radiator, so the coolant coming down will be cool in the same to later take the low hose. The cool coolant passes through the water pump, often located behind the radiator fan. The water pump, in its role, it will circulate the coolant throughout the system. Similar to the function performed by the AC compressor circulating the refrigerant in the whole system. Finally, the cold coolant enters the engine again. Becomes hot to repeat the journey.

(Fig. 47) -Water pump.

(Fig. 48) -Thermostat.

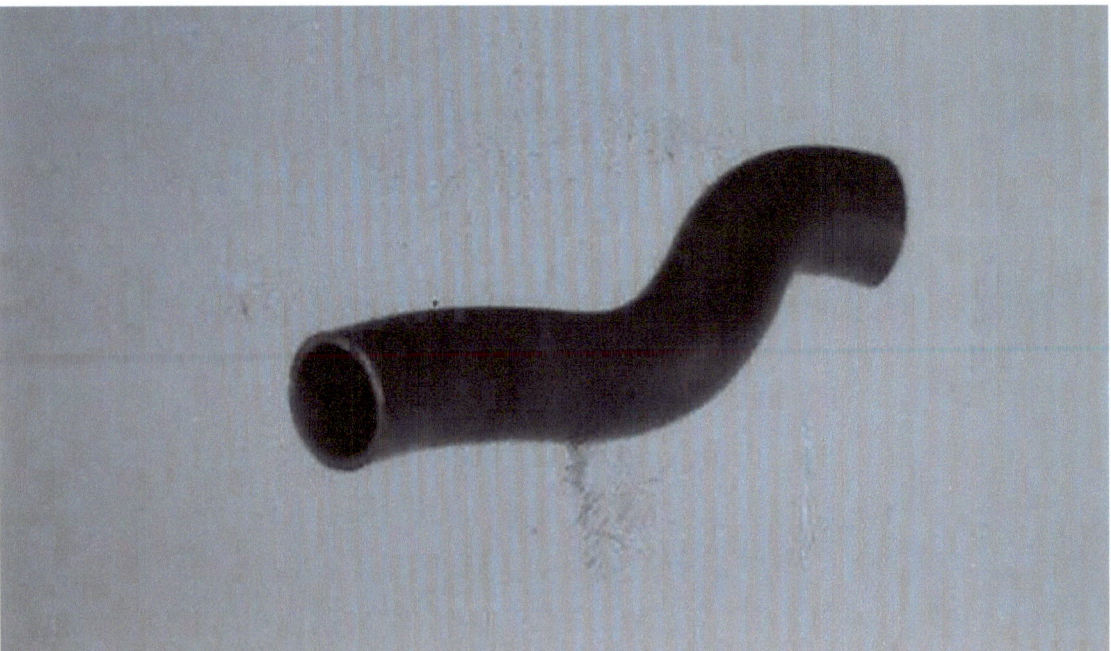

(Fig. 49) -Radiator hose. Are two, the first one enters the top. And the second one exits the bottom.

(Fig. 50) -Radiator.

Never attempt to remove the radiator cap with the engine warm. The coolant will come out violent.

(Fig. 51) Radiator fan.

(Fig. 52) -Recovery tank.

TWO SERIOUS PROBLEMS:

I- Clock engine temperature, marking a higher reading than normal.

II- The radiator has launched an explosion of hot coolant, prevail intense smoke.

POSSIBLE CAUSES:

a) There is maybe a leak of coolant or may have a low level of coolant in the system.

-You must check the coolant level in the recovery tank. Once the engine cools, checks the level of coolant in the radiator. Added if needed.

- If the problem is due to a leak and this loss of coolant is by a hose connection, tighten the hose clamp.

- If the water pump which has a leak, it must be replaced.

- If is leaking the heater core, there will be some shadow of grease inside the car. The carpet passenger will be moist. As a temporary repair adds "Stop-leak product" to the radiator. You will need to find and repair the connection leaking. Later more detail about the heater core.

b) Maybe the heater core is locked. It is failing.

- Turn on the heating in a high position. The heater fan must be also in high position. The reason is to eliminate heat from the engine. Parked in a safe place, put your car in neutral. Run engine in order to increase circulation of coolant. If the heater core is extremely locked, may be needed first rinse the cooling system. Then, rinse the heater core. If the obstruction is irreparable, there is not the best option for replacing the component.

c) It can be that the radiator fan is not working.

- Inspect fan operation. It is good to keep in mind that the radiator fan starts to work when the engine reaches a predetermined temperature.

- Check the connector feeding with 12 volts the radiator fan.

- Maybe you need to replace the radiator fan.

d) Maybe fault the water pump. This device takes coolant from the bottom of the radiator and pumped through the engine. If the pump is failing, replace it.

- Check the water pump pulley. A pulley failing can cause vibration under the hood.

e) Maybe the radiator has a leak.

- Add "radiator stop-leak". If the not result, go to the store and buy a new one.

REPLACING THE RADIATOR.

1- It is time to drain the radiator. To drain means to lower water or liquid through an opening in order to leave empty the container. Place a bucket underneath to catch the antifreeze. There are two ways to drain the radiator:

A- Disconnect the low hose.

B- Open petcock: it is a valve at the bottom of the radiator. It opens like a faucet in the kitchen.

2- While the system is empty from the coolant, remove the parts that could block the radiator access. It can be the thick air inlet pipe. This thick pipe is secured at each terminal with a clamp. Loosen the clamps.

3- You may need to remove the radiator fan.

4- Remove the clamp of high hose. Remove the hose from radiator and then, hang it with a piece of wire face-up on a site of the engine.

5- If the vehicle has an automatic transmission, there may be transmission lines further connected to the radiator. You must untie these lines and clear the radiator area.

6- Remove the screws that secure the radiator to its position in the automobile.

7- Remove the hose that is in the radiator cap. (In some cars, the radiator cap it appears on the right with a thin hose that goes up to the recovery tank). A pressure using a clamp tool can release said hose.

8- Remove the lower radiator hose.

9- With screws and hoses removed. And with the fan removed, the radiator can be lifted out of the vehicle.

(Fig. 53) -A view from the top of the radiator. Note the thin hose to the recovery tank. Note also, the upper radiator hose.

FLUSHING THE COOLING SYSTEM

1- Locate the input line to the heater core, long enough to accommodate a T.

2- Drain the radiator.

3- Using a knife, cut the inlet hose to the heater core.

4- Swipe its clamps on each end of the cut hose. Insert the T.

5- Tighten the screws of these clamps.

6- Tie a nozzle (something like a short tube-shaped elbow) to the mouth of the radiator. Put a bucket in front of the nozzle. This is in order to collect water that rinses the system.

7- To flush the cooling system, you can use water from a faucet with a normal hose gardening. Connect the hose to the T.

8- Open the faucet and start to flush.

9- Once you finish, keep the T installed for future rinses. Do not forget to put firm the lid to prevent coolant out of the system.

10- Drain the whole water. Remove the nozzle from the mouth of the radiator.

11- Fill the radiator with antifreeze/ coolant and put the lid.

HEATING SYSTEM consists of the following elements:

1- The heater core.

2- Blower fan.

3- Heater core hoses.

CYCLE

The automotive heater cycle has nothing to do with the AC system. On the heater system is used antifreeze/coolant, something very different to the AC system, which is using refrigerant R134a. The coolant coming from the engine continues until the heater core: this heater component is mounted in a housing. That is, the heater core and the evaporator they are usually located together under the instrument panel, both components are enclosed in the same box. As same the evaporator is used to supply cool passengers, the heater core is used to keep them warm. Both as a tool taking the effect of a fan that emits air over the winding of one or the other.

THE MOST COMMON PROBLEM:

(The heater It does not work)

POSSIBLE CAUSES:

a) There may be a low level of coolant in the vehicle.

- Check the coolant level in the recovery tank and radiator. Remember to wait for the engine cool.

b) The heater core may have some obstruction.

- Flush the heater core.

- Replace the heater core.

c) The heater core fan is not working.

- Check the fan.

d) The heating switch in the control box for AC and heating is failing.

- Check the switch.

e) Maybe the thermostat is failing.

- The coolant never gets hot enough to provide good heating. In some cases, the answer is replacing the thermostat.

For the engine car works as it should, the coolant must achieve a temperature of about 90° C. When this liquid is heated, the thermostat will open and will enable a flow into the radiator. On its exterior, the thermostat usually indicates at which temperature opens. However, it may that the thermostat no longer performs their function (open and close). And as this element has expired, it may end open or close indefinitely. This will bring serious problems, more serious than a person could imagine.

For example: If the thermostat remains closed: There will be no flow of coolant to the radiator. With dry radiator, the engine will overheat to finish melting. And otherwise, if the thermostat ends open, with to coolant to radiator it will lower the temperature of the engine. Thus the engine will end abraded.

REPLACING THE THERMOSTAT

In many vehicles, the thermostat is located where the top of the upper hose radiator connected to the engine. If the thermostat is off, there is a good reason for the heating in the passenger area may not work properly. One thermostat opens not provide a comfortable temperature in cold weather.

Removing the thermostat some coolant will be discarded, so it is best to drain the radiator to a level belonging to the position of the thermostat.

1- A little drain radiator.

2- Using a socket to loosen the screws of the sheath thermostat.

3- When you remove the casing, you may need a screwdriver to remove the thermostat instead.

4- The gasket can be a ring tied around the thermostat itself. Or can be a sheet separated to the wrapper. Clean the old gasket or install the new thermostat with a gasket identical to the above.

5- Using the socket tighten the screws.

6- Run the engine and check for a leak of coolant.

(Fig. 54) -Radiator high hoses it is coming from the engine top where is hiding the thermostat.

(Fig. 55) Heater core.

The automobile and its good air conditioning

REPLACING THE HEATER CORE.

1- Drain the cooling system.

2- Disconnect the heater core hoses in the firewall. The firewall is the wall that separates the engine area of the passenger cabin. Refers to firewall face looking to the engine area.

3- Remove the instrument panel.

4- Remove the screws on the housing where is hiding the heater core.

5- Open said housing and take the heater core.

(Fig. 56) -Scheme of automotive heating and cooling system.

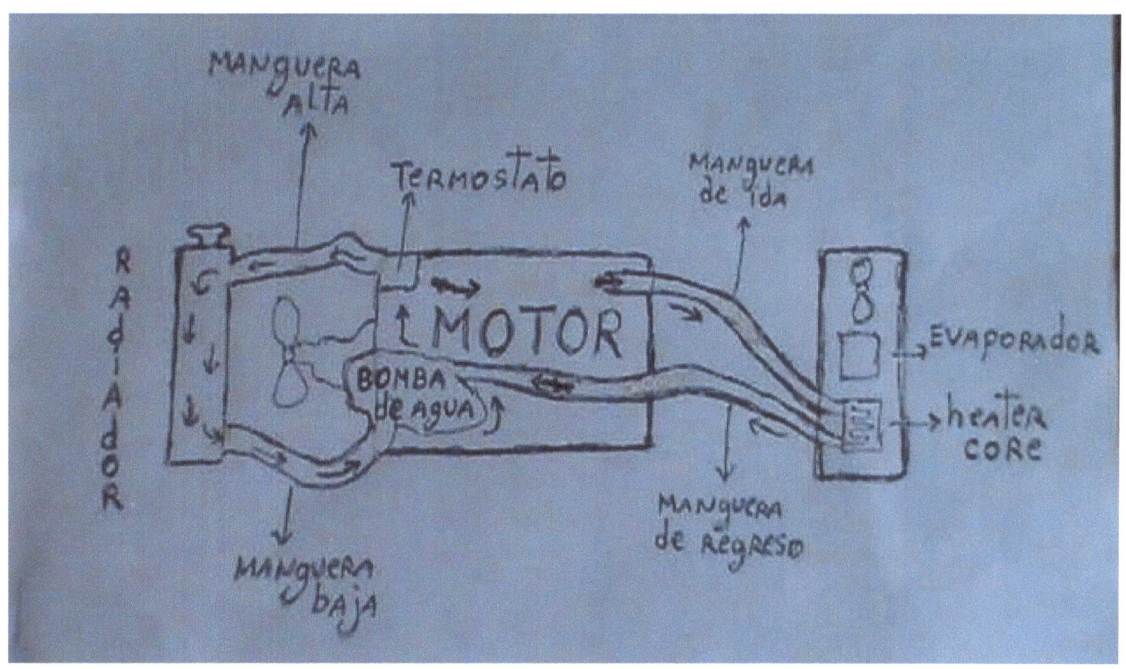

And so far this wonderful magic of air-conditioning of the automobile and its heating system. I hope that this information serves you something and... get to work. It's been a pleasure having your attention to this book which, as I said at the beginning, I do not intend to make it a rule, but a representation or similarity to which it could face you if you decide to undertake the arrangement of your car.

It was a work written by the young elder.

www.ingramcontent.com/pod-product-compliance
Lightning Source LLC
Chambersburg PA
CBHW051156220526
45473CB00003B/791